ALL ABOARD READING™

Level 3
Grades 2-3

APOLLO 13
Space Race

A13

By Gail Herman

Screenplay by
William Broyles, Jr. & Al Reinert
and John Sayles

With photos from NASA and
the movie *Apollo 13*

Grosset & Dunlap • New York

Just a Movie?

Three astronauts are in space. Two days ago, they blasted off from Earth. Two days from now, they will land on the moon.

So far it has been a smooth flight. An easy flight. The astronauts float lazily in their spacecraft. They talk and joke.

Then . . .

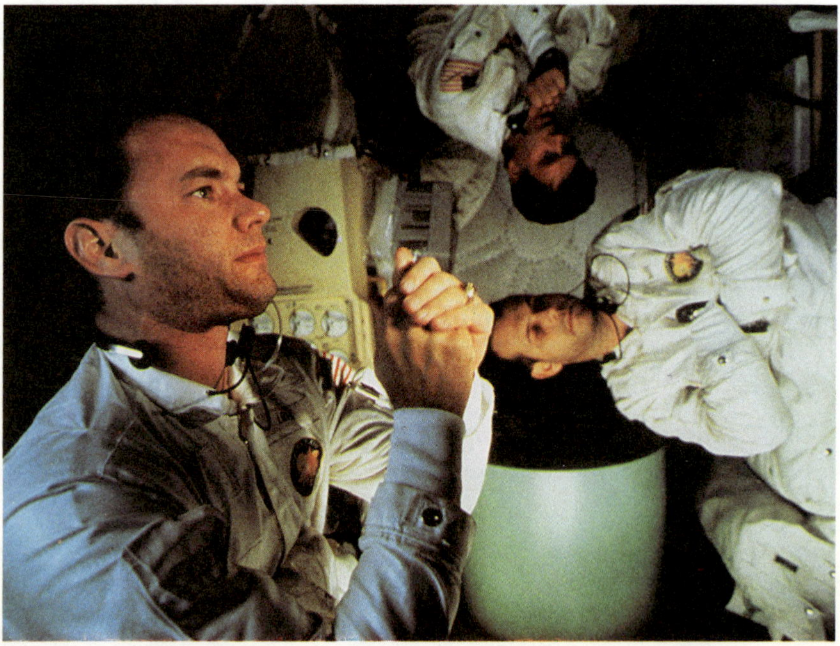

Put Beginning Readers on the Right Track with ALL ABOARD READING™

The All Aboard Reading series is especially for beginning readers. Written by noted authors and illustrated in full color, these are books that children really and truly *want* to read—books to excite their imagination, tickle their funny bone, expand their interests, and support their feelings. With three different reading levels, All Aboard Reading lets you choose which books are most appropriate for your children and their growing abilities.

Level 1—for Preschool through First Grade Children
Level 1 books have very few lines per page, very large type, easy words, lots of repetition, and pictures with visual "cues" to help children figure out the words on the page.

Level 2—for First Grade to Third Grade Children
Level 2 books are printed in slightly smaller type than Level 1 books. The stories are more complex, but there is still lots of repetition in the text and many pictures. The sentences are quite simple and are broken up into short lines to make reading easier.

Level 3—for Second Grade through Third Grade Children
Level 3 books have considerably longer texts, use harder words and more complicated sentences.

All Aboard for happy reading!

Photo credits: pp. 13, 16, 19, 21, 23, 25, 27, 29, 31, 32, 36, 42, 47, 48, courtesy of NASA; p. 15, AP/Wide World Photos.

Published by Grosset & Dunlap, Inc., a member of The Putnam & Grosset Group, New York. ALL ABOARD READING is a trademark of The Putnam & Grosset Group. GROSSET & DUNLAP is a trademark of Grosset & Dunlap, Inc. Published simultaneously in Canada. Printed in the U.S.A.

Library of Congress Cataloging-in-Publication Data

Herman, Gail, 1959-
Apollo 13 : space race / by Gail Herman ; screenplay by William Broyles, Jr. & Al Reinert and John Sayles ; with photos from NASA and the movie Apollo 13.
p. cm. — (All aboard reading)
"Level 3, Grades 2-3."
1. Project Apollo (U.S.)—History—Juvenile literature. 2. Space race—United States—Juvenile literature. 3. Space vehicle accidents—United States—Juvenile literature. 4. Apollo 13 (Spacecraft)—Accidents—Juvenile literature. [1. Apollo 13 (Spacecraft) 2. Project Apollo (U.S.) 3. Space vehicle accidents.] I. Title. II. Series.
TL789.8.U6H53913 1995
629.45'4'0973—dc20 95-13070
CIP
ISBN 0-448-41117-2 A B C D E F G H I J AC

BANG! WHOOSH! An explosion rips through the spacecraft. Walls shake. Warning lights flash. An alarm sounds.

The craft bucks like a wild horse. It spins and dives in every direction. The commander grabs the controls and the wild bucking stops.

"We have a problem," he radios to Mission Control in Houston, Texas.

The commander's voice is calm. But he knows they are in trouble. Deep trouble.

Before, the spacecraft had two full oxygen tanks. Now, one is empty. The other is draining fast. Soon there will be no oxygen for the men to breathe. There will be no oxygen to help fuel the craft and keep it going. The men will be

trapped in space. They will die.

Does this sound like a scene from a movie? An exciting space adventure movie? It is. The movie is called Apollo 13.

But Apollo 13 was a real spacecraft. There was an explosion. And three astronauts—Commander Jim Lovell and crewmates, Fred Haise and Jack Swigert—really did face death.

Mission Control back on Earth had a plan to save them. Get the astronauts out of the main part of the spacecraft, the command module, Odyssey. Move them into the LEM, the lunar module.

The LEM was a much smaller spacecraft attached to the command module. It was designed to land on the moon.

But there could be no moon landing. Not anymore. The LEM had an even more important job—to save the astronauts' lives.

Could it? The tiny spacecraft has oxygen, electricity, and power. But it was built to keep two men alive for just two

THE APOLLO SPACECRAFT

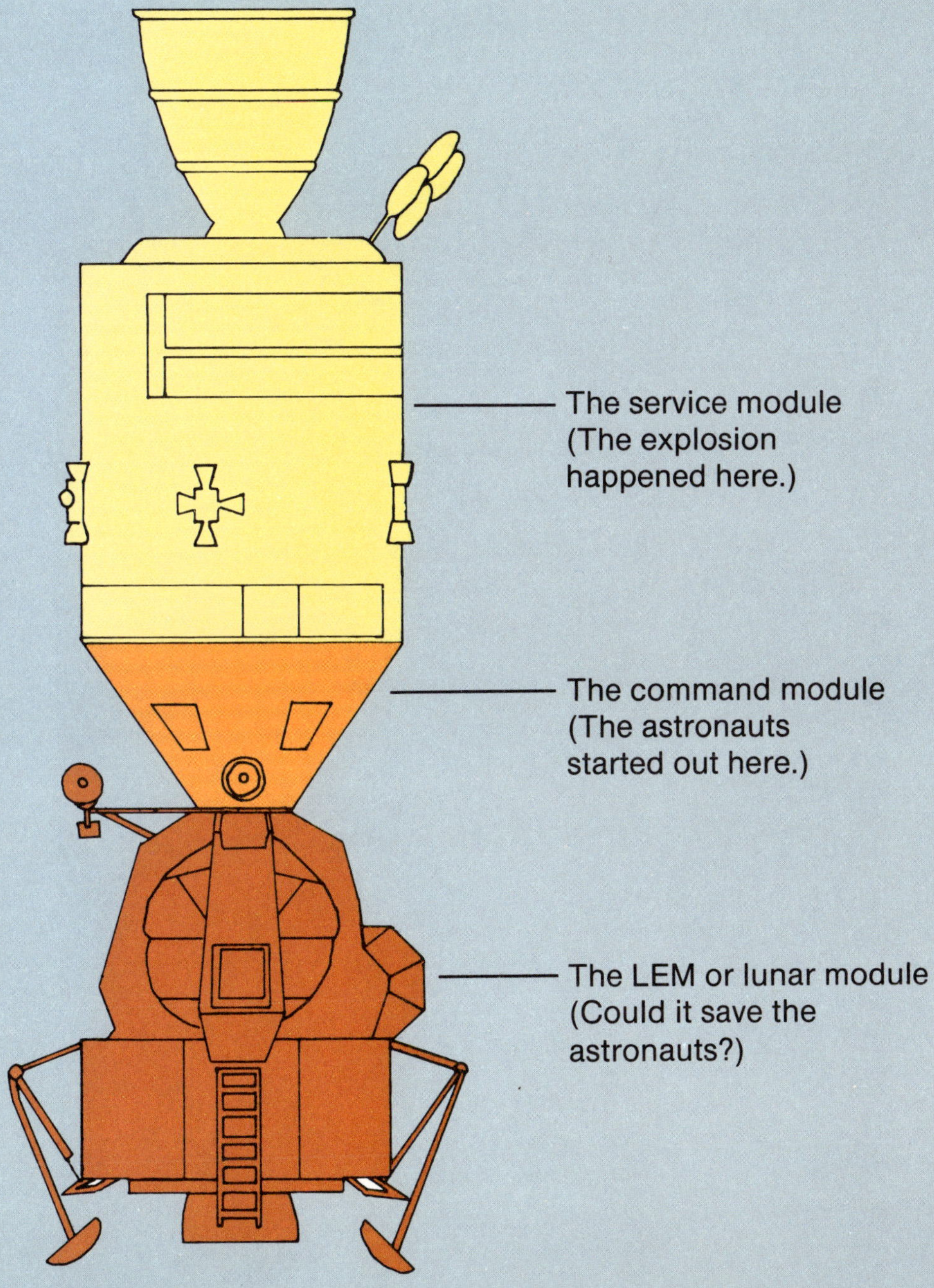

days. There are three men. It will take them four days to get back.

Three astronauts. Four days. How will they survive?

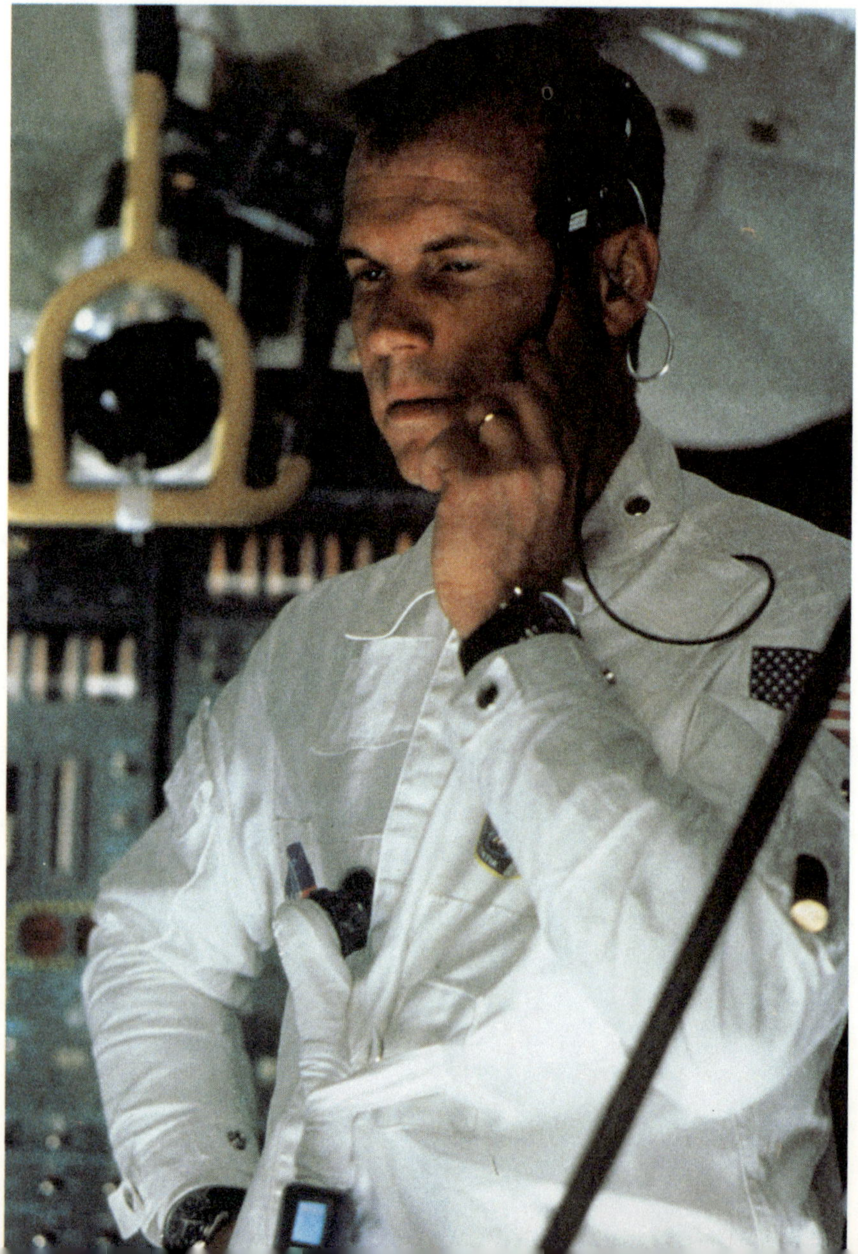

Get Ready . . .

The news is flashed around the world. Will the astronauts die? People everywhere are worried. Even the Soviet Union—the nation America sees as its worst enemy—wants to help.

Today there is no Soviet Union. It is broken up into Russia and lots of other smaller countries. But then it was very big. And very strong. The Soviet Union and the United States were the superpowers of the world. And they didn't get along.

Each country saw the other as a threat. Each believed war could break out without warning. Bombs could rain down. Entire cities could be wiped out.

Scientists were put to work inventing new weapons. If the Soviet Union built a big bomb, America wanted a bigger one. If America had a new rocket, so did the Soviets.

Then, on October 4, 1957, the Soviet Union surprised the world. They sent a satellite into space. A 180-pound, small metal ball. For the first time ever, an object made by human beings traveled around Earth. It circled—or orbited—the planet every 95 minutes. The satellite was called Sputnik.

The United States was worried. The Russians had started to explore space. Soon they could control space. And whoever controlled space could control the world.

The United States had to keep up. They had to catch up. People thought it

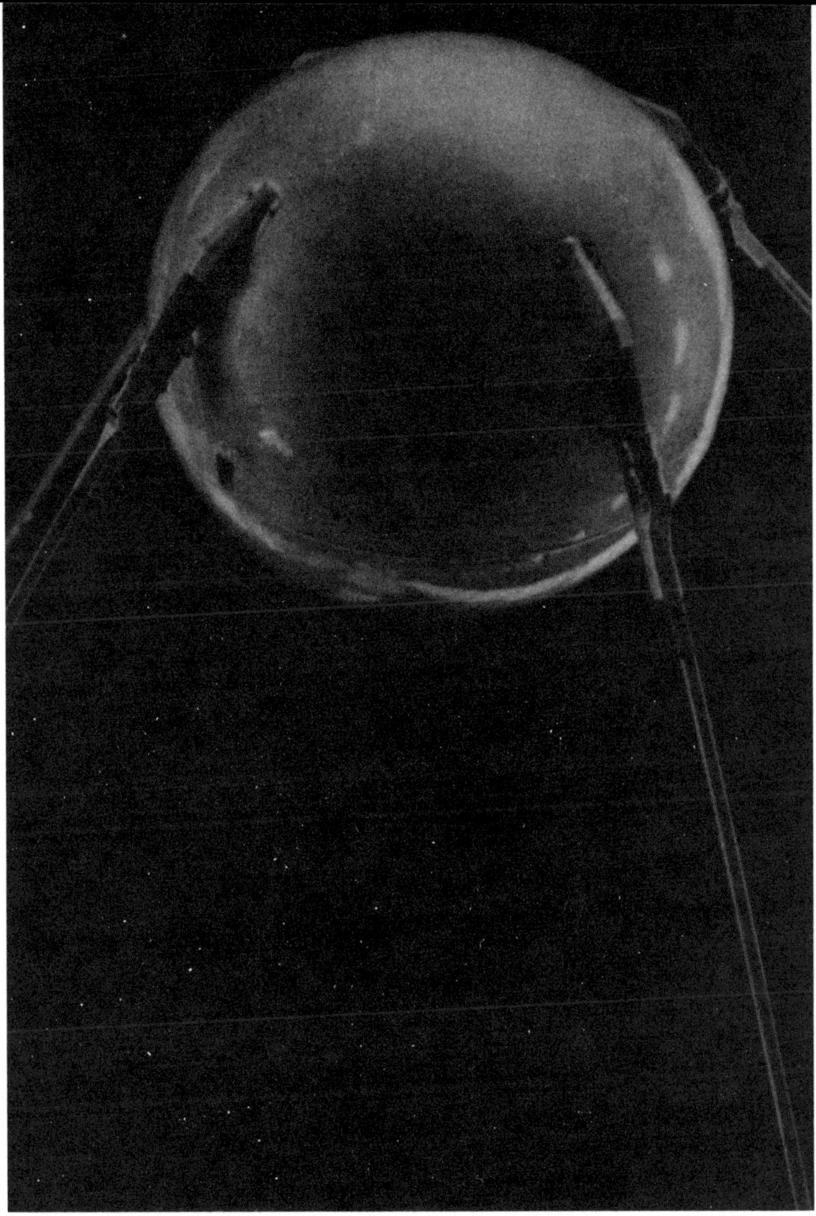

Sputnik was about the size of a basketball.

was a contest that the country could not afford to lose.

The space race had begun!

Get Set . . .

Back in the 1950s, the idea of space travel was very scary. It meant going into the unknown. Traveling into a silent, airless, black-as-night place. Would astronauts be able to eat? Sleep? Breathe? Go to the bathroom?

Could people really travel in spaceships—and come back alive?

The Russians decided to find out. Just one month after Sputnik, they launched Sputnik II. This satellite carried a passenger. But the passenger wasn't a person. Too risky, scientists thought. Instead, they sent a dog named Laika.

Laika ate food and drank water from

special containers. Scientists checked their readouts. Laika stayed alive!

But scientists had a problem. They did not know how to bring Laika back. Sadly, she died after a few days in space.

Meanwhile, the United States was trying to launch satellites, too.

On their first try in late 1957, the rocket fired. It hovered above the launchpad. Millions of Americans held their breath. Would it zoom into space? No. The rocket exploded on the launch site.

The United States tried again. And again. Then on January 31, 1958 the first U.S. satellite made it into space.

Still the Russians kept one step ahead. They sent two more dogs into space—Strelka and Belka. This time they knew how to bring the dogs back. Strelka and Belka dropped to Earth by

The first U.S. satellite was about 6 feet tall.

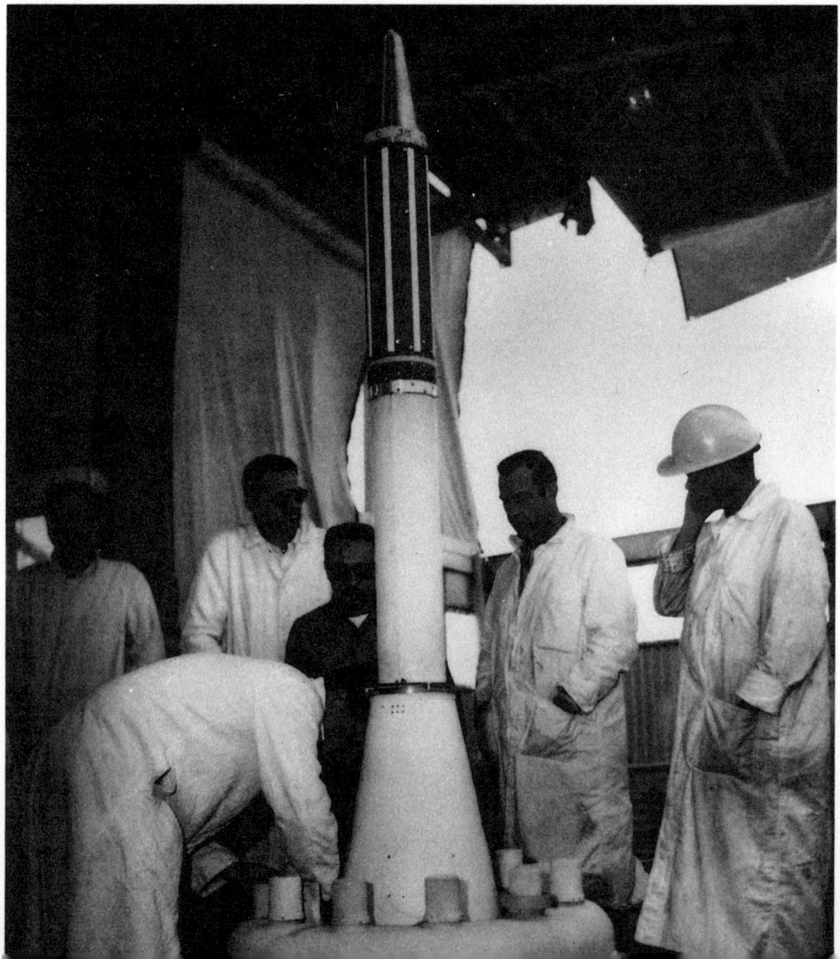

parachute—and they were just fine!

Next, the Russians did the unthinkable. It was 1961. Just three years after Sputnik. They trained a cosmonaut—a Russian astronaut—and sent him into space aboard Vostok I.

The cosmonaut was a pilot named Yuri Gagarin. (You say it like this: gah-GAR-in.) His tiny spacecraft rose almost 200 miles in the air. Then it went into Earth's orbit and circled the planet. The flight lasted 108 minutes.

Did Gagarin survive the trip? Yes! The Russians had done it! They'd sent a man into space.

The Russian leader Nikita Khrushchev (You say it like this: KROOSH-cheff) thanked Gagarin. "Now let other countries try and catch us," he dared the world.

In the United States, people were amazed and worried. Could they catch up now? Was America ready to send a man into space? The truth was U.S. scientists weren't sure. So they decided to send a chimpanzee named Ham. First Ham had to train for the trip. He learned how to pull certain levers when different lights flashed.

On January 31, 1961 Ham rode in a spacecraft for 17 minutes and 5,000 miles. But he had trouble coming back. The engine wasn't working right.

Ships waited for Ham to splash down at the target spot in the Atlantic Ocean. They searched the sky. No sign of Ham or his spacecraft.

But 130 miles away, Ham had already splashed down. He was safe, and nobody knew it. Ham bobbed in the water for hours before he was rescued.

If a chimp could do it, a man could, too. The United States felt ready to send a man into space.

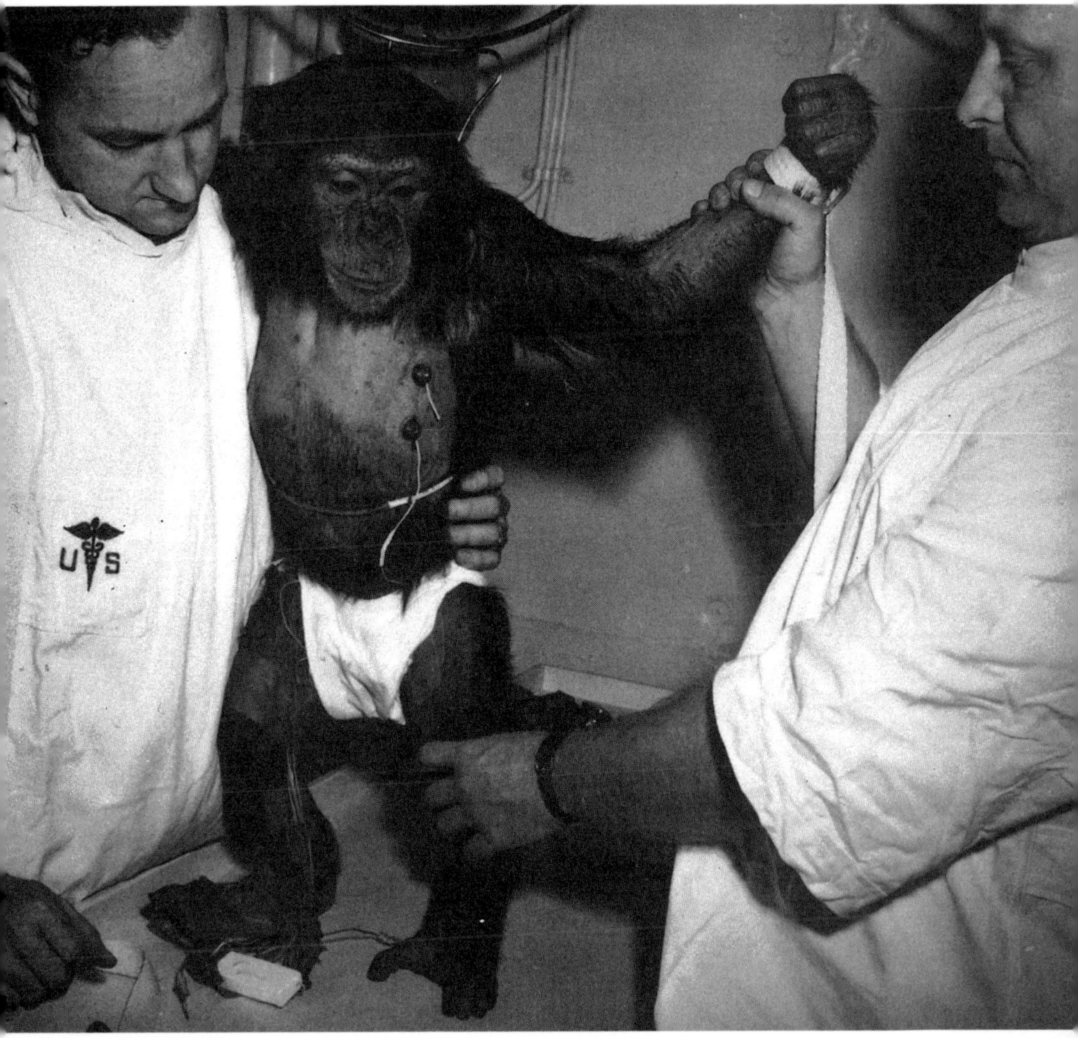

One month later, astronaut Alan Shepard waited in his spacecraft for lift-off. It was a long wait. Mission Control kept running tests—thousands of them. Shepard stayed strapped in his seat for hours. He had no room to move. The craft was the size of a phone booth!

At last, everything was ready. Alan Shepard shot up more than 100 miles above Earth. "What a beautiful view!" he radioed to Mission Control. He did not go into orbit as Gagarin did. But he came back safely.

People cheered. There were parades and parties. And the President of the United States, John F. Kennedy, made an important speech. He set a goal. Before 1970 an American would walk on the moon.

People couldn't believe it. Shepard had been in space for just a few minutes. A trip to the moon would take days. It called for a different rocket. A new super-powerful rocket. It seemed impossible.

The first American in space

But Americans couldn't stop thinking about the Russians. They had the first satellite. The first animal in space. And the first human. Wasn't it America's turn to lead the space race?

Go!

America took a giant step to the moon on February 20, 1962. Friendship 7 blasted off. The astronaut, John Glenn, planned to orbit Earth seven times.

First orbit . . . second orbit . . . third orbit. Everything was going fine. But all of a sudden, warning lights flashed.

The heat shield was loose! If it fell off, Glenn would be in great danger. When he dropped back to Earth, his craft would heat up hundreds of degrees. No one could live through that!

But Glenn couldn't stay in space forever. He had no choice. He had to try and land and just hope the heat shield didn't fall off.

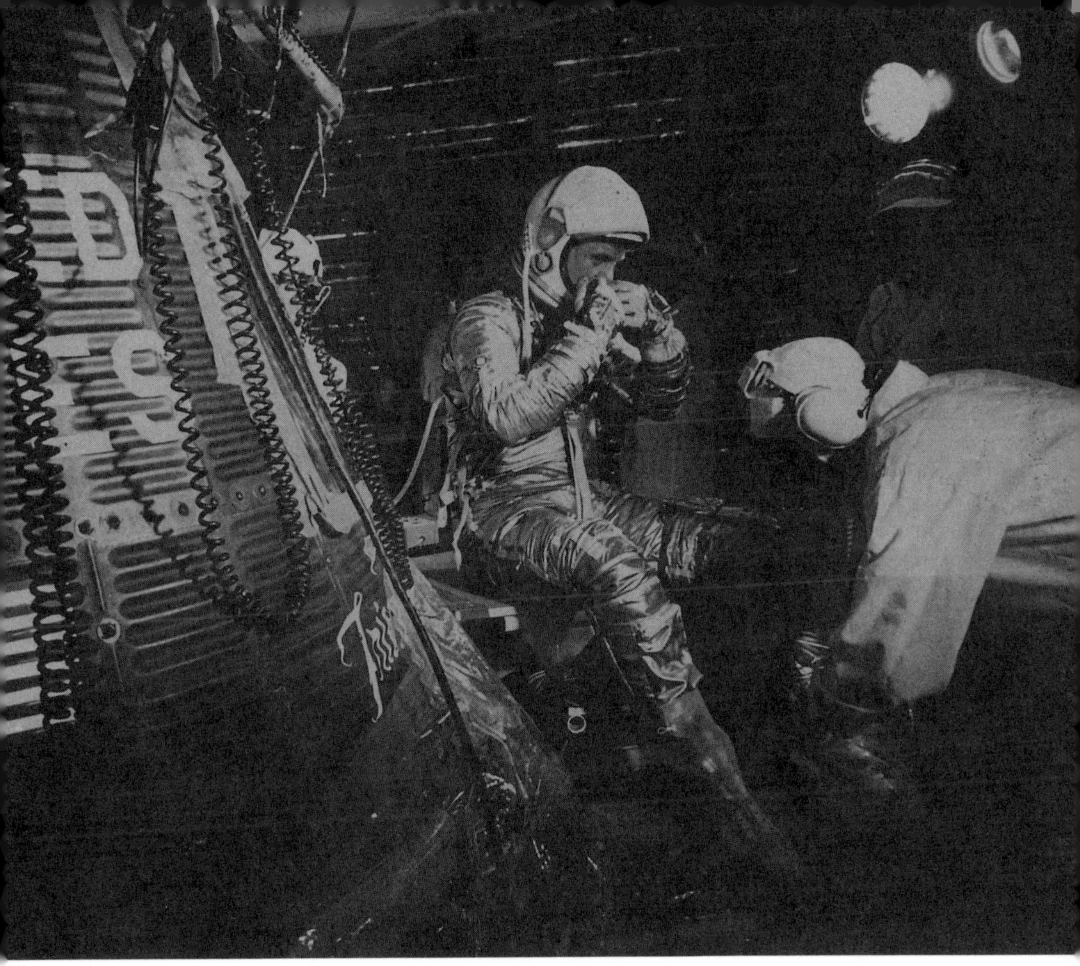

Getting John Glenn ready for his trip

Glenn was ready. The spacecraft dropped. Thump! The craft rattled. Parts fell off. Big flaming chunks of spacecraft hit the window. Glenn shook with the force. But the spacecraft slowed down and gently splashed down in the Atlantic Ocean.

Later scientists discovered the heat shield wasn't loose at all. The signal had been wrong.

John Glenn was a hero. He had stayed in space for five hours. He had risked his life. Because of him, the United States was catching up with the Russians.

But the Russians still had more "firsts." In 1963 they sent the first woman into space. She stayed there for three days. And in March 1965, a cosmonaut actually left his craft and floated in space. It was the first space walk!

America, however, wasn't far behind. An astronaut, Ed White, tried it just weeks later. He walked in space while his spacecraft traveled 18,000 miles per hour!

These space walks proved something

important. With the help of a space suit, astronauts could leave their craft. They could walk on the moon!

Ed White "walks" in space.

There were more space flights. Astronauts stayed in space for up to two weeks at a time. By 1966 the United States had lots of information. They had a new super-strong rocket. And they had a brand-new space program. Scientists called it Apollo.

Space fever hit like never before. Everybody caught it. The Apollo missions would put people on the moon!

But the program began with disaster. On January 27, 1967, Apollo 1 sat on top of its Saturn rocket going through a countdown.

This was just a test launch. The spacecraft wasn't even leaving the ground. But three astronauts were inside—Gus Grissom, Ed White, and Roger Chaffee.

Apollo 1 after the fire

Suddenly flames shot out of the windows of Apollo 1. Ground crews raced to put out the fire. But it was too late. The three astronauts were dead.

What went wrong?

Scientists took apart the entire Apollo spacecraft to find out. They studied it piece by piece. They made over 1,000 changes. They tested and retested.

Two years passed. Finally, they were ready to start again. Apollo 7 was launched on October 11, 1968. The crew stayed in space for eleven days. And they came home safely without one single problem.

Apollo 8 was going to test the LEM in Earth's orbit. But then the United States received a secret report. The Russians were planning to fly up to the moon and circle it. They would be first—again!

Quickly, the Apollo 8 mission changed. Its astronauts would circle the moon. America would be the first!

Christmas day, 1968. The three Apollo 8 astronauts became the first people—ever!—to travel around the moon. The first to see its far side—the side that doesn't face Earth. Apollo 8 orbited the moon ten times. And the astronauts took photos to bring back home.

One of the first pictures of Earth rising over the moon

Then it was time for the big one—Apollo 11. The mission that would land astronauts on the moon. All the astronauts wanted to go. But only three men could be first. Neil Armstrong, Edwin "Buzz" Aldrin, and Michael Collins were the lucky ones.

July 16, 1969. Lift-off! An ear-splitting blast launched the Apollo 11 spacecraft. The ground trembled. And the astronauts streaked up, up, up. Faster and faster. On their way to the moon.

The trip to the moon took three days. Then, while the world held its breath, Armstrong and Aldrin moved into the LEM, the Eagle. Gently, the Eagle separated from the command module. It dropped lower and lower until it came to a stop.

"The Eagle has landed," Armstrong announced.

Armstrong and Aldrin were on the moon!

The United States was way ahead in the space race now. The Russians stopped trying for a moon landing altogether. They didn't want to be second. But in November 1969, Americans went to the moon again with Apollo 12.

Four months later, the United States launched Apollo 13.

Apollo 13 in Space

It is Monday night on Apollo 13—56 hours after launch. It is just after the explosion. The engine in the service module was destroyed. The astronauts get ready to move into the LEM.

Mission Control tells them to "power down" the command module. Everything that uses power, electricity, or fuel must be turned off for now. They'll need to turn it back on later—if they can.

Jim and Fred crawl into the LEM. Jack stays in the command module. Quickly he turns off dial after dial. He is nervous. Every second counts.

At last the power-down is complete. Jack squeezes into the LEM with Jim and Fred. It is a tight fit. They can barely turn around. But, for the moment, the astronauts can rest.

At Mission Control, scientists are hard at work. There is so much to figure out! How can they stretch the LEM's water, oxygen, and fuel?

Everything must last long enough to save the astronauts' lives.

Then Mission Control has an idea. A good idea. Let gravity do the work. Let the moon's gravity pull the spacecraft around the moon. Fire the LEM's rocket engine to increase speed. Then Earth's strong gravity will take over to pull Apollo 13 toward Earth. It will be like a free ride. And it will use less fuel.

It is the only choice.

April 14th, 1:43 A.M. The astronauts orbit the moon. On the far side, they lose radio contact with Earth for 28 minutes. It is a scary time. Mission Control can't speak to them. They can't help if there is a problem.

Finally, the astronauts hear Mission Control again. They've made it around the moon. The astronauts fire the engine once to speed up the craft and send it in the right direction. Back to Earth.

Now the astronauts power down the LEM. Instrument panels go black. There is no heat. No cabin lights. The astronauts are cold. The temperature is a little above freezing. And already, they are tired.

They take turns sleeping in the command module. But it is hard to rest. The command module is even colder than the LEM.

Suddenly there is another setback. Mission Control is getting a dangerous readout. The problem: bad air.

All people breathe out a gas called carbon dioxide. Usually, nature cleans the air. But inside the LEM, carbon dioxide is building up. It is poisoning the air.

The problem must be fixed or the astronauts will die.

The LEM has special filters to clean the air. Now they are almost used up. The command module has filters, too, but they are the wrong shape.

Somehow the astronauts must find a way to make the command module filters work in the LEM, using only what they have on board!

Mission Control radios up a plan. The astronauts rip cardboard from their

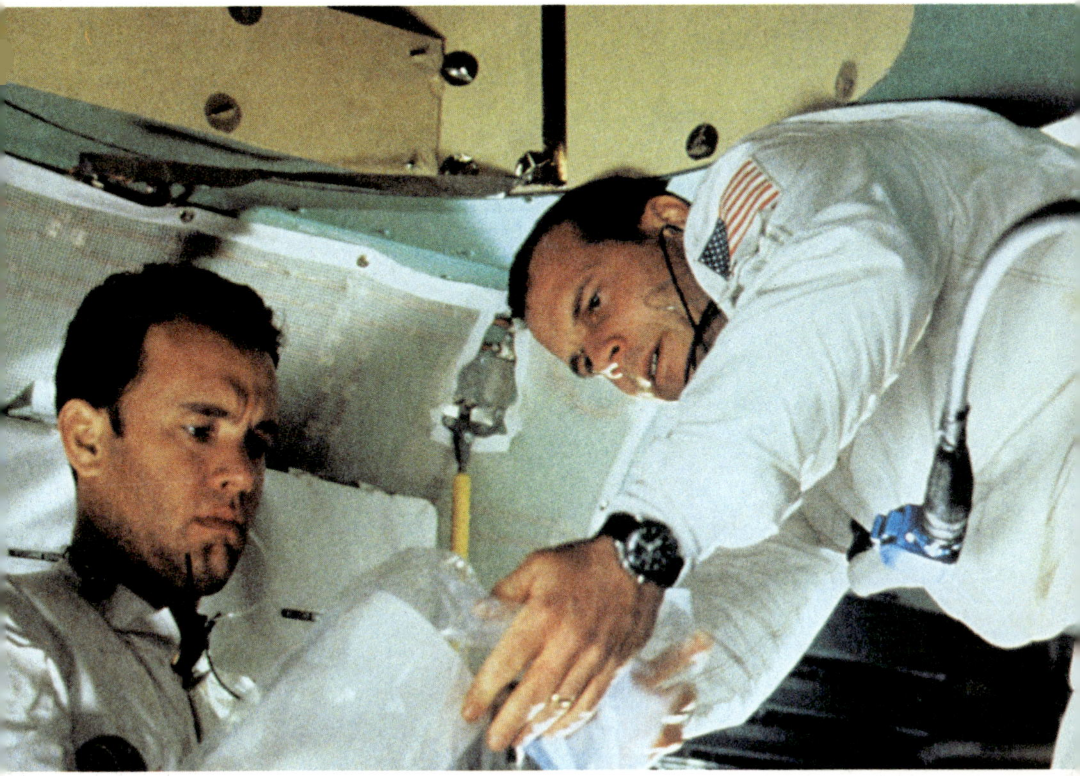

flight plan booklet. They grab tape and a sock. They cut plastic from a package of long underwear.

They put new filters together. And they work!

The air is healthy again. For now, the astronauts are safe.

Apollo 13: Coming Home

For four days now, the astronauts have been crowded into the LEM. Windows and glass are fogged with cold. Fred is sick. He has fever and chills. But the astronauts are almost home.

It is almost time to turn the command module back on, system by system—just the way they turned it off.

The astronauts go over every step. Everything must be done perfectly. Or all will be lost.

"You're go," says Mission Control. On board Odyssey, Jack throws the first switch.

Hours later, the power-up is complete. The astronauts leave the LEM for good and move back to the command module.

Jack flips another switch. It separates the LEM from the Odyssey. They do not need the LEM anymore. It drifts off into space. The men take a last look at the tiny spacecraft that has saved their lives—so far.

It is the final hour. The command module drops to Earth. The astronauts see only blackness out their window. Are they heading in the right direction? They can't be sure. They have no way of checking.

Finally they see a curved shadow. Hanging over it is a bright white moon. The curved shadow is Earth. They are heading home!

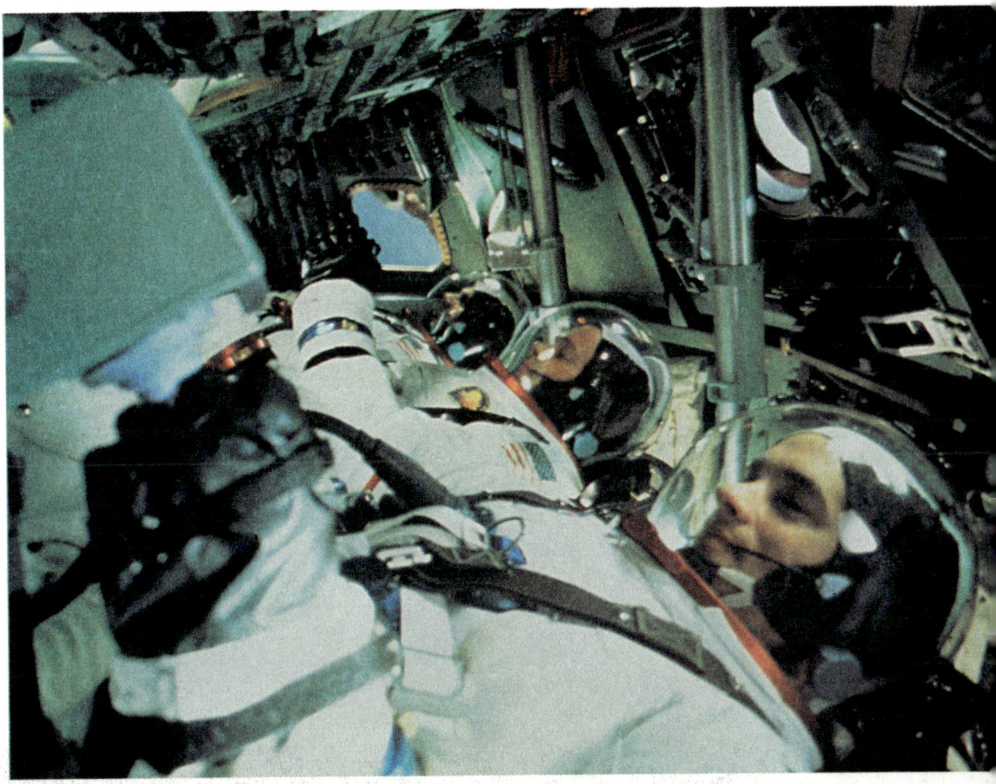

Boom! The spacecraft rams into Earth's atmosphere—the air all around the planet. The astronauts lose contact with Mission Control.

Outside, black space turns pink then orange then red. Tiny fiery sparks streak by the window. The ship shakes wildly.

At Mission Control, minutes tick by like hours. They radio to the spacecraft. Are the astronauts okay? There is no answer.

"Houston standing by," they call again. And again. Nothing.

Finally there is static. Then a voice. It's Jack! They made it!

By now it is Friday morning. Under bright blue skies, Jim opens the parachutes. The craft slows to 20 miles per hour. Thirty seconds later, it slices smoothly into the South Pacific Ocean.

The astronauts are home. Safe and sound.

The Last Apollo

The flight of Apollo 13 had a happy ending. But no one could forget what had happened. It made people think. Should the space program go on? Maybe it was too dangerous. Too costly—in money and in lives.

Plans were already set for other missions. So four more Apollo spacecraft landed on the moon.

In 1972 the Apollo program ended. People thought that was it. No more Apollo missions. Ever. But they were wrong. Three years later, everything was dusted off for another run.

Apollo-Soyuz had a special name and a special mission. An Apollo spacecraft

Apollo links up with Soyuz in space.

linked up with a Russian craft. The two spacecraft stayed in Earth's orbit. They did not go to the moon. They did not make new discoveries.

Yet this mission was a first for both Russians and Americans. For the first time, astronauts and cosmonauts talked together and worked together in space.

The space race was really over. People had begun to feel that space belonged to everybody. But space exploration was not finished.

Mars—the red planet

Today space shuttles still blast off from Earth. And space probes—robot crafts without people on board—fly to different planets. They've gone to Mars, Venus, Jupiter, and as far as Neptune.

New space programs may take us back to the moon. They may take us much, much further than the moon. And just think—it all began with a race. A space race.